THE THEORY OF EVOLUTION IS A RESULT OF ERRONEOUS EXTRAPOLATION

JUNE MEIMBAN

ISBN 978-1-64458-340-1 (paperback)
ISBN 978-1-64458-341-8 (digital)

Copyright © 2018 by June Meimban

All rights reserved. No part of this publication may be reproduced, distributed, or transmitted in any form or by any means, including photocopying, recording, or other electronic or mechanical methods without the prior written permission of the publisher. For permission requests, solicit the publisher via the address below.

Christian Faith Publishing, Inc.
832 Park Avenue
Meadville, PA 16335
www.christianfaithpublishing.com

Printed in the United States of America

For since the creation of the world God's invisible qualities—His eternal power and divine nature—have been clearly seen, being understood from what has been made, so that people are without excuse.
—Romans 1:20 (NIV)

CONTENTS

Introduction 7

Chapter 1: Critical Thinking 9

Chapter 2: Adaptation Does Not Prove Evolution 19

Chapter 3: Mathematics, DNA, and Nanomachines 30

Chapter 4: God Exists 54

References 59

INTRODUCTION

I will begin with a quote from John Ruskin. He said, "The work of science is to substitute facts for appearances, and demonstrations for impressions." What does this have to do with the theory of evolution? The theory of evolution is based on the hypothesis that species randomly mutate and by natural selection, the fittest organisms survive, and their new traits are passed on to the next generations. Then over long periods of time, these small changes are accumulated in their genetic material, and species gradually are transformed to other species. Evolutionists, scientists who adhere to this theory, claim that the theory of evolution means that humans descended from apes, and life appeared on earth by pure random chance. They further say that fossil records, DNA analysis, and even mathematical simulations all support the validity of the theory of evolution.

When you were first taught about the theory of evolution in school, did you even question its validity? You probably believed it and still do, right? Perhaps it's because you were required by your school to learn it, and you have to take tests on it.

This book will walk you through a step-by-step and easy-to-understand process, where I will substitute facts for appearances and demonstrations for impressions and help you understand why the theory of evolution is wrong.

The view presented here is my own and does not represent any organization.

CHAPTER 1

Critical Thinking

Consider the following data derived from a 1991 record of flight arrivals. In 1987, the Department of Transportation required all US airlines to report data on whether the arrival is late or on time. For now, let's look at the comparison between Alaska Airlines and America West.

1991 Data from Five (5) major Airports

Airline	No. of Arrivals	ON-TIME Arrivals	LATE Arrivals	% Late
Alaska Airlines	3,775	3,274	501	13.3%
America West	7,225	6,438	787	10.9%

Given this table, which airline do you think has the worse performance? Most people would conclude and say, "The probability of Alaska Airline being late, as indicated by the percent Late column, is higher, so it has the worse performance." Let's see if you're right.

Take a look at the more detailed tables below. The data is now broken down by airport.

Alaska Airline	ON-TIME Arrivals	LATE Arrivals	Total Arrivals	% Late
Los Angeles	497	62	559	11.1%
Phoenix	221	12	233	5.2%
San Diego	212	20	232	8.6%
San Francisco	503	102	605	16.9%
Seattle	1,841	305	2,146	14.2%
Total	3,274	501	3,775	13.3%

America West	ON-TIME Arrivals	LATE Arrivals	Total Arrivals	%Late
Los Angeles	694	117	811	14.4%
Phoenix	4,840	415	5,255	7.9%
San Diego	383	65	448	14.5%
San Francisco	320	129	449	28.7%
Seattle	201	61	262	23.3%
Total	6,438	787	7,225	10.9%

Now what can you notice? America West has the worse performance than Alaska Airlines on every airport. It's the reverse of what you derived from the earlier table not broken down by airport. As you can notice, America West flies mostly out of Phoenix which has sunny days (most of the time) while Alaska Airlines flies mostly out of Seattle which experiences more rainy and cloudy days.

The above is a classic example of Simpson's paradox. There are many real-life examples of this occurrence whenever data are aggregated. The conclusion you might get is the reverse of what could be derived when you look at the data in more detail. Actual data in test scores, school admission rates, sports, etc., have been used as classic examples as well. Simpson's paradox is named after Edward Simpson who wrote about it in 1951. Although a British statistician, G. Udny Yule, first described it in early 1900s.

THE THEORY OF EVOLUTION IS A RESULT OF ERRONEOUS EXTRAPOLATION

Making conclusions—the right conclusions—is not as easy as one might think. It requires critical thinking, attention to detail, understanding of the data, and many other things.

Consider this next example.

Given that x and y are any real numbers, and y is a function of x, i.e., y = f(x). I know the function f(x) but you don't. I want you to figure it out.

The only clue I will give you is a table of *sample* values of x and the corresponding values of y below. Now, can you determine the function f(x)?

x	y
14.04	1.00
1.67	1.00
1.57	1.00
-11.00	1.00
76.97	1.00
-538.78	1.00

If you answered yes, it is f(x) = 1, you are wrong. Your hypothesis is wrong. As you can see below, the function y = sin(x) also fits the data.

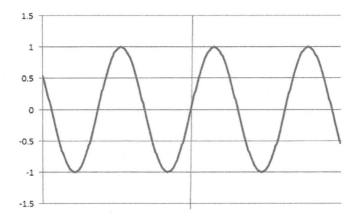

The point I am making here is even if your hypothesis fits all the given data points, it doesn't mean that your hypothesis is correct. There are many other possible hypotheses that will fit the data. This is a key concept you have to understand. Evolutionists would claim that "all" the observed data fit the predictions from the theory of evolution.

First, they provide you of the only data that fit their theory. There are many data points that do not fit their theory. I will touch on those in later parts of this book.

Second, even if the data points they provide fit their theory, it does not follow that their theory is correct.

Let me repeat one sentence in the preceding paragraph. Evolutionists would claim that "all" the observed data fit the predictions from the theory of evolution. Now given this, they would conclude that their theory must be correct. Whatever happened to "substituting facts for appearances and demonstrations for impressions"?

Evolutionists would point to gradual changes in fossil records. They would say, "Look, this fossil of an antelope-like animal has short neck. Then we found another one. We carbon-dated it, and we detected that it lived millions of years after the first one we found, and its neck is longer." Then they will conclude that the giraffe "evolved" this way.

First, evolutionists assume that the giraffes' long necks "evolved" to help them feed. They call this the high-feeding hypothesis. I'm putting quotes around "evolve" because evolutionists are assuming that their theory is true. So whenever they write their scientific articles, they use the word "evolve" as if it was true.

Now, they're finding out that the high-feeding hypothesis is weak. There are parts in Africa where giraffes really like to eat by reaching up to the top of trees. But there are also parts in Africa where even when food is scarce, the giraffes don't reach up. In the July 7, 2010, issue of *Zoologger*, Michael Marshall mentioned a new hypothesis. Biologists are now saying that the giraffe's long neck is a result of sexual selection. Male giraffes fight for females using their necks, swinging it against the other male giraffes, as if in a duel. So

THE THEORY OF EVOLUTION IS A RESULT OF ERRONEOUS EXTRAPOLATION

those with long necks win. They survive. They mate and pass on that trait to their offspring.

Whoa! Wait! Stop!

Show us the proof! Don't *extrapolate*!

For a neck to significantly "grow," according to evolutionists, there has to be a series of mutations over long periods of time. I put quotes around "grow" because evolutionists assume the necks "grew" through gradual series of mutations. They say this is "proven" by fossil records. Whoa! Wait! Stop!

Do you even know the chance of a mutation? Most mutations are bad, like sickle cells. Instead of surviving, you die. Random mutations are rare. And even by *lucky chance* of a good mutation leading gradually to the desired outcome (e.g., longer neck), it could take millions and millions of *lucky chances*.

By all probability, it is most likely not possible due to the complexity of the DNA. Again, the giraffe thing is a data point that they claim fits their theory. But it doesn't follow that their theory is correct! We will get back to the time element and mutation rate later. It requires some basic probability theory.

Perhaps, after more data is observed in the future, they, the evolutionists, would determine that their sexual selection hypothesis is weak. They will come up with new hypotheses for sure. I got one for them. How about, well, the giraffes have long necks because that is how they were created. And all the animal fossils that the scientists have seen that have varying neck lengths? Those animals were also created that way. It just seemed to their naked eyes that there were "gradual changes" since the necks have varying lengths. They arranged the fossils in carbon dating-produced times, and they "correlate" the necks' length pattern with the passage of time. In statistics, we know that correlation does not imply causation.

The evolutionists simply extrapolated again! But take note, the creation hypothesis also fits the data.

Now you see, as in the $y = f(x)$ example, two possible hypotheses fit the observed data.

In order to see why correlation does not imply causation, we can look at this example.

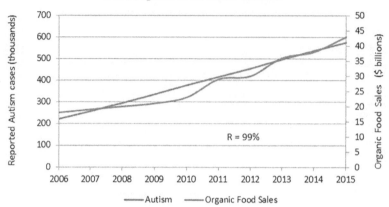

Does organic food cause autism?

Sources: https://nces.ed.gov/fastfacts , Organic Trade Association Surveys , US Dept. of Education

The correlation between organic food annual sales and the number of autism is very high in this example. The R-value is a measure of how strong the linear relationship is between two variables. In this case, it is almost 100 percent. One might conclude that organic food causes autism, which we know is erroneous. *Correlation does not imply causality!*

In order to show causation, controlled experiments are needed. This is what's being done in drug research for example. Subjects are divided into groups, and different groups receive different treatments. Experiments are carefully designed to control for the effects of various variables and treatments. Outcomes are measured for statistically significant lift. Even this method does not guarantee that cause and effect can be established. Nevertheless, this is more powerful than a simple correlation study.

Conclusions that seem to imply cause and effect (causality) based on correlation are widely found in scientific research that claims evolution is true. They simply extrapolate correlation into causality, which is erroneous.

THE THEORY OF EVOLUTION IS A RESULT OF ERRONEOUS EXTRAPOLATION

Now, let's look at how evolutionists fail to distinguish between correlation and causation.

Assume a hypothetical data below. It is a fossil record of measurements of giraffe-like animals gathered by a scientist.

Age of fossil (in million years)	length of neck (in feet)
5	2
4	3
3	4
2	5

If you'll compute the R-value here, it will be very high. So yes, there is very high correlation. You can even fit it into an equation. But does that imply causation? Meaning that time (age) is the cause of the necks of giraffes getting longer? Not necessarily! Again, the evolutionists are extrapolating! In this case, they are assuming that correlation implies causality.

In this case, the existence of fossils, even though there are differences and when arranged in chronological order shows a pattern of increasing lengths over time, is not a proof of the validity of the theory of evolution.

These really old fossils simply mean that those creatures lived on earth in the past. In fact, no transitional forms have been found in the fossil records. All the fossils ever found are fully formed functional distinct species. The Cambrian explosion, an event that paleontologists say happened 541 million years ago, show that almost all the animal phyla appeared fully formed and already in complex forms during this time. There is no evidence of gradual intermediate change from one species to another. What evolutionists have been doing is "forcing" their erroneous argument by arranging the fossils in their "assumed" historical classification, i.e., cladistic tree, to "prove" a common ancestry and gradual transformations. But this is a circular and subjective argument and not based on scientific cause-and-effect studies. They assume evolution to prove evolution. Now, evolutionists have come up with new hypotheses such as punctuated equilibrium which proposes that once a fossil appears it becomes stable for long periods. This is just another speculation! Evolutionists are desperate in forcing their arguments. In fact, the more they dig into fossil records, the more they find that there are no transitional forms.

Stephen Jay Gould, an evolutionary paleontologist admitted, "The absence of fossil evidence for intermediary stages between major transitions in organic design, indeed our inability, even in our imagination, to construct functional intermediaries in many cases, has been a persistent and nagging problem for gradualistic accounts of evolution." (Stephen Jay Gould, "Is a new general theory of evolution emerging?" Paleobiology (1980), 119-130.)

In fact, the dating of the fossils is questionable and highly controversial; and radiometric dating is flawed.

In 2005, Mary Schweitzer, a scientist at North Carolina State University, found soft tissues and blood vessels in a fossilized leg of a T. Rex showing that dinosaurs walked on earth not millions of years ago but mere thousands of years ago. See http://history.com/news/scientists-find-soft-tissue-in-75-million-year-old-dinosaur-bones and https://youtu.be/B4cOY-dp1D0 throwing the fossil dates into question.

Another controversy surrounding the dating of fossils is the circular argument around it. Fossils are dated based on the rocks they're found in but the rocks are dated based on the "assumed" chronological arrangements of the fossils.

J.E. O'Rourke, in "Pragmatism versus Materialism in Stratigraphy," American Journal of Science, Vol 276, January 1976, p.47, wrote "The intelligent layman has long suspected circular reasoning in the use of rocks to date fossils and fossils to date rocks. The geologist has never bothered to think of a good reply, feeling the explanations are not worth the trouble as long as the work brings results."

So, what's happening here is evolutionists are forcing the issue by using this circular technique of using rocks to date fossils and fossils to date rocks then use correlation to "establish" causality! This is not science! Circular arguments lead to bogus results and correlation does not prove causality.

THE THEORY OF EVOLUTION IS A RESULT OF ERRONEOUS EXTRAPOLATION

The "invented" geologic column you find in textbooks is simply a product of evolutionists' imagination and not based on scientific facts. The so-called index fossils simply are erroneous fabrications. For example, *graptolites* are the index fossils for 410 million year old rocks (Earth Magazine September 1993), however, these were found alive in South Pacific in 1993. The *lobe-finned fish* is also used as the index fossils for 325-410 million year old rocks but is found alive today in the Indian Ocean.

In fact, as evolutionists dig more fossils, the more they find that their theory is wrong and the fossils point to evidence of a catastrophic global flood. Numerous examples of embedded vertical petrified trees have been found in the fossil record cutting supposedly millions of years in the geologic column. Some are even upside down. This is an evolutionist's nightmare because it points to a scientific proof of Noah's flood. As the strong flood waters dislodge these trees the sediments begin forming layers of rock cementing the trees together with the creatures. (John Morris, The Young Earth Master Books, Green Forest, AR, 1994 pp 93-117).

Archaeologists have been finding Goliath-like human remains all over the world. This is consistent with the bible (Genesis 6:4) but contrary to evolutionary theory. See for instance this video: https://youtu.be/lN5t0p6TtYo. Archaeologists are also unearthing ancient advanced civilizations all over the world wiped out by a catastrophic global flood challenging the mainstream view of human history. Watch https://youtu.be/4VlM1ar_LlE

Dinosaur "toe-print" fossils are being discovered all over the world such as in China in 2013 and in Australia. These toe-prints capture just the tip of the toes and evidence show the dinosaurs were being held up by rising water and they were trying to escape by running up the mountains. This is best explained by the catastrophic global flood. (http://www.icr.org/article/dinosaurs-swimming-out-necessity)

In fact, the Cambrian explosion is best explained by the biblical flood! (https://answersingenesis.org/fossils/fossil-record/one-lifes-unexpected-explosion).

You've been lied to by evolutionists. They've used fraudulent evidences and hoaxes to "prove" their theory. A perfect example is Ernst

Haeckle's embryos still being found in modern biology textbooks. The purported drawings of embryo sequences claimed to show that humans share a common ancestry with various animals by looking at their embryo's similarities. These drawings have been proven fakes and forgeries. Haeckle had been charged and convicted by university court for over 100 years and yet his drawings are still used in many textbooks today. Among them are Evolutionary Biology (3rd edition Sunderland, MA: Sinauer Associates 1998) by Douglas Futuyma and Molecular Biology of the Cell (3rd edition) by Bruce Alberts, et al.

The Piltdown Man is another example of a hoax used by evolutionists. In 1953, almost forty years after it was claimed to have been discovered as the "missing link" between man and apes, it was exposed as a forgery. The skull was from a modern man and the jawbone and teeth were from an orangutan. They've been fabricated to look old. A complete lie!

The Horse evolution fraud is another example. The "evolution" of the horse timeline is a misleading arrangement of fossils using evolutionary assumptions (circular argument) and yet are being found in modern-day textbooks. The list goes on and on – Nebraska Man, Java Man, Neanderthal Man.

https://evolutionnews.org/2015/04/haeckels_fraudu/
https://evolutionisntscience.wordpress.com/evolution-frauds/

Do not extrapolate. If you believe in evolution because you've seen data such as the above fossil record data, I'm sorry to tell you, my friend, you've been brainwashed.

You will see in the rest of this book many examples of how evolutionists have erroneously concluded causality when in fact only correlation is observed—from fossil records, apparent similarities in organisms, phylogeny studies which simply establish correlation and not cause and effect.

In fact, phylogeny studies are circular arguments, i.e., they assume that evolution is true to prove evolution! You might be shaking your head by now, right? Put on your critical thinking cap and read on!

CHAPTER 2

Adaptation Does Not Prove Evolution

Evolutionists would point to the observed ability of species to adapt to their environment as a "proof" of evolution. They cite examples such as the peppered moth, *Biston betularia*—a light-colored moth in eighteenth century England.

Before the industrial revolution, this moth was able to elude predators such as birds because its light color somewhat camouflaged it by living in light-colored trees. When factories and pollution took over, the bark of the trees turned darker. The moth became an easier target. Scientists observed that a dark variety of the moth became more dominant. They would explain that the light-colored moths died off, the dark-colored ones survived and passed on that trait to the next generations. Then evolutionists would say that's a proof of survival of the fittest aspect of evolution; therefore, the theory of evolution is correct.

Whoa! Whoa! Stop! Stop!

The evolutionists are extrapolating again!

The dark-colored moth is still a moth. Even after a million years, it will still be a moth. Maybe it will have a different color, wing size, faster speed of flight perhaps, but it will still be a moth. Yes, adaptation is real, but do not extrapolate it to mean evolution is real. The moth will never turn to another animal. Observed small changes in species such as the color of moths and the length of beaks in finches are simply examples of adaptation and not proof of transformation of one species to another.

Scientists have tried to prove that the theory of evolution is correct by growing cultures of bacteria in laboratories and observing it for long periods of time—decades, that's bacterial time. In humans, that could be equivalent to millions of years. Because humans can have children say as early as twenty years, but bacteria can multiply in a matter of thirty minutes.

Now they claim they have observed new bacterial species they've never seen before. Then they'll conclude it is a proof of evolution. Whoa! Wait! Stop! They are still bacteria!

The main reason scientists miss this difference between adaptation and evolution is because of brainwashing. They've been brainwashed by schools who teach that adaptation is a proof of evolution.

Pause here and think about this really hard because you've been brainwashed too. Get that out of your system, it is an erroneous extrapolation.

Brainwashing is a result too of the loose use and interpretation of the word "nature." When people say or hear "it's natural" or "it's part of nature," they subconsciously think that things just happen by chance because it is just the "way it is"—not designed, not planned, and so on—because they see it happening every day. So when they see something that's part of an animal or a plant, they'll say, "It's part of nature. It just happens that way!" They never bother to think critically, investigate, and think outside the box. And the bad part is that the word "nature" or "natural" was used by Darwin in his theory. So people are "brainwashed" into believing that Darwin's theory is real because it's natural due to his use of the phrase "natural selection."

In mathematics, there are things you can prove without any shadow of doubt called laws or theorems, such as the law of large numbers, Pythagorean theorem, Fermat's last theorem, Poincare's theorem. As a side note, last time I checked, Goldbach's conjecture is still not proven, and the Riemann hypothesis is still a hypothesis. People might say it's just semantics . . . Ah, this is deep, but I will get back to this topic later.

Let's get back to adaptation.

THE THEORY OF EVOLUTION IS A RESULT OF ERRONEOUS EXTRAPOLATION

Adaptation is real, but evolution is not. Organisms can adapt and adapt as long as they want, but that is adapting. Not in a million years will they "evolve" to another organism.

Here is another bad extrapolation that is used by evolutionists. When bacteria multiply and develop antibiotic resistance, evolutionists say this is a proof of evolution. Wow, that is adaptation, not evolution.

Most bacteria have around 3 million ($3*10^6$) base pairs. Think of a base pair as a step in the DNA ladder. The DNA is like a code that determines how a sequence of amino acids are formed then linked (this is called the primary structure), then folded (secondary) then folded again into 3-dimensional protein structures (tertiary and quaternary). The last folding steps are very complex. There are 20 different amino acids used by living organisms. The proteins' shape and characteristics determine a lot of things. In humans, for example, the color of your eyes, skin, susceptibility to diseases, etc., are results of the proteins produced by your DNA. Protein synthesis is one of the key areas currently being researched in nanotechnology.

Variation within the same species are pretty small. In humans for example, it's been estimated that there is a variation in human population DNA of about 0.10 percent and that accounts for the differences we see in the population—different skin color, height, bone structure, etc.

Some single cell organisms have 500,000 base pairs in their DNA. Humans have about 3,000,000,000 (3 billion). Frogs have about 3 billion too. Sharks, 3 billion; Newt, 15 billion; Lilly, 100,000,000,000. That's right, 100 billion.

DNA has a unique double-helix structure. A turn in a strand has about 10 base pairs.

In bacteria, for instance, structural proteins are used in the development of its membrane—its "skin" or cell wall.

Without the membrane or the "skin," the bacteria is exposed! In fact, its internal organs would simply disperse. The membrane protects it from its outside environment. In fact, many drug research on antibiotics are based on how to attack this membrane. Penicillin kills bacteria by inhibiting its ability to build the cell wall itself.

In a given bacterial culture, variation in the organisms' characteristics, the membrane in particular, exists already. Some may have membranes that are more resistant to drugs, some that are not. So in a clinical test of a drug for killing bacteria for example, the more resistant bacteria survive, those nonresistant ones die.

Yes, this is survival of the fittest. It's real. The survivors now are able to pass on that trait to the next generation. So the culture that survives has more drug resistance. Bacteria can multiply very fast. Now, when there is mutation (in bacteria, it's about 2×10^{-8} chance per base pair replication event), the mutation gives higher chance of having more variability or variety. But they are still the same—bacteria. Not another organism. Mutation increases variability. It does not produce a new animal.

Do not extrapolate adaptation into evolution! One kind of animal will never turn to another kind of animal. Apes will never turn to humans and never did.

Here is another erroneous extrapolation. Kids in schools are being taught that human and chimpanzee DNA are 99 percent similar. Then teachers extrapolate this and say, "See? We evolved from chimpanzees!"

Whoa! Wait! Stop!

You're brainwashing the kids! First of all, it is *not* 99 percent. It is more like 70 percent. And *similarity does not prove evolution!*

The 99 percent figure is way off the mark. It was based on a 1975 research by Mary-Claire King and A.C. Wilson comparing the primary structure of proteins between humans and chimpanzees. With proteins being made from amino acids, which are made from reading the DNA code, they made a "big leap of inference" (extrapolated!) in saying that human DNA and chimpanzee DNA are 99 percent similar. We now know more advanced things about the DNA—its sequence in particular.

Recent studies based on looking at the patterns in the DNA show the difference is more like 25–30 percent. So the correct statement is, "Human and chimpanzee DNA are only 70–75 percent similar." Now, these are based on recent technology. As technology improves we might find more dissimilarity.

THE THEORY OF EVOLUTION IS A RESULT OF ERRONEOUS EXTRAPOLATION

Humans have 23 pairs of chromosomes. Chimpanzees, orangutans, and gorillas have 24. Again, teachers would say, "Look, they're similar. We humans must have evolved from apes."

Whoa! Wait! Stop!

That's wrong extrapolation! Similarity does not prove evolution!

But the damage has been done. Kids have been brainwashed. And most people have been brainwashed. They've been brainwashed by the educational system that based its decisions on biased opinions, bad science, erroneous extrapolations, misunderstanding correlation versus causality, perhaps knowingly or unknowingly.

I believe that the theory of evolution is the biggest hoax in human history. It is the biggest blunder of science.

Yeah, most scientists are intelligent people. But intelligent people sometimes make mistakes.

In later part of this book, you will see how some of the best minds in mathematics got stumped by a simple probability problem and got it wrong, big time. You might be thinking of how can the best mathematicians be stumped by a simple probability problem?

To read about it now, check this out: http://marilynvossavant.com/game-show-problem.

You have to realize that the DNA was discovered in 1953, about 100 years after Darwin came up with the theory of evolution in 1859. Darwin had no idea that cells, when seen under powerful microscopes, are very complex.

In fact, he wrote,

> If it could be demonstrated that complex organs existed which could not possibly have formed by numerous, successive, slight modifications, my theory would absolutely breakdown. (Charles Darwin, *Origin of Species*, 1st ed. p. 189)

In 1996, Michael Behe, a Lehigh University biochemistry professor, showed the world a microscopic organ—the flagellum—that is "irreducibly complex." He wrote a book titled *Darwin's Blackbox:*

The Biochemical Challenge to Evolution, published by Free Press. For a background, watch the video found in www.revolutionarybehe.com

The flagellum is like a tail that a single-cell bacteria, like E.coli uses to swim and move around. Under high-powered microscopes, scientists have discovered that this little, tiny organ is actually a molecular machine. In other words, a nanomachine. It is composed of 40 different protein components that look like parts of a modern motor. It is so efficient and complex and is powered by proton motor force, two gears that can go forward and reverse, operates at 17,000–100,000 rpm. It is so advanced that some scientists say, "It is the most advanced machine in the universe."

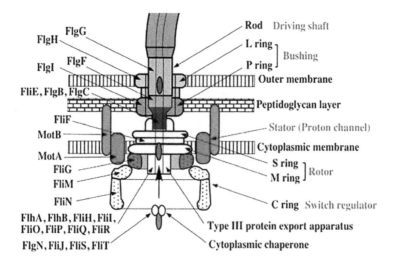

The flagellum diagram. (Source: https://evolutionnews.org/2011/03/michael_behe_hasnt_been_refute/By Jonathan M.)

The "irreducibly" term in "irreducibly complex" is very important. I say the "complex" part is good enough to debunk the theory of evolution!

The "irreducibly" part refers to the fact that if any of these parts is removed, the whole structure does not work. It is all or nothing. All the parts have to be present and assembled in microscopic space with precision far more advanced than what we currently have in

THE THEORY OF EVOLUTION IS A RESULT
OF ERRONEOUS EXTRAPOLATION

nanotechnology. In fact, many new research efforts in nanotechnology are based on "reverse engineering"—the flagellum. Now, the theory of evolution breaks down because for this "irreducibly complex" machine to "gradually evolve," the parts must be assembled gradually over millions of years. And that's not possible because it is irreducible.

Evolutionists now would point to other flagella in other bacteria that have "similar" structure with less parts. But those are irreducibly complex nanomachines as well! (Read on!) Some evolutionists would argue that the flagellum is not irreducible—that they can take a few parts and the bacteria could still swim, albeit not as efficiently.

Whoa! Wait! Stop! Stop extrapolating!

So evolutionists are saying that even if the bacteria's flagellum had some missing parts, over millions of years, it "knew" how to manufacture the missing parts in the future and put them in the right precise locations? One evolutionist even said the protein parts were floating around already inside the bacteria and by series of random mutations, they were put in place. So the bacteria had missing parts, swimming with handicap (what happened to survival of the fittest?) and by some lucky mutation the missing part appeared? These are their counterattacks to Michael Behe's "irreducibly complex" argument.

This is another example of really bad extrapolation by the evolutionists. For now, I will give a background on how even the nonscientists, i.e., lawyers, got involved in this scientific debate.

Michael Behe's demonstration of the irreducible complexity of the flagellum is a strong argument against the fundamental assumptions of the theory of evolution. Like I mentioned above, the "complex" part is good enough . . . more on this later. But for now, let's focus on "irreducibly complex."

In a landmark case in 2005, *Kitzmiller et al. v. the Dover School District*, Behe was an expert witness for the intelligent design (ID) argument. The details of this case are quite extensive, so I will just summarize it here for easier reading.

Basically, the theory of evolution has been taught in public schools as the explanation of the origin of life. Standardized tests in schools implicitly required students to learn this theory. The students were therefore "forced" to learn it. Beginning in 2002, William (Bill) Buckingham and Alan Bonsell, members of the Dover Area School District Board of Education, advocated teaching intelligent design alongside evolution. At a board meeting on June 7, 2004, Buckingham objected to the proposed use of the textbook *Biology* by Miller and Levine, which is based on the theory of evolution, saying it was "inexcusable to have a book that says man descended from apes with nothing to counterbalance it."

So basically the issue was whether to include creationism (or intelligent design) alongside evolution in the education of students. This seems like a fair compromise, right? At least, this gives students a chance to learn two competing theories about the origin of life.

So on November 19, 2004, the Dover Area School District issued a press release stating that beginning in January 2005, teachers would be required to read a statement to the ninth grade biology class at Dover High School. The statement is this:

> The Pennsylvania Academic Standards require students to learn about Darwin's theory of evolution and eventually to take a standardized test of which evolution is a part.
>
> Because Darwin's Theory is a theory, it is still being tested as new evidence is discovered. The Theory is not a fact. Gaps in the Theory exist for which there is no evidence. A theory is defined as a well-tested explanation that unifies a broad range of observations.
>
> Intelligent design is an explanation of the origin of life that differs from Darwin's view. The reference book *Of Pandas and People* is available for students to see if they would like to explore this view in an effort to gain an understanding of what intelligent design actually involves.
>
> As is true with any theory, students are encouraged to keep an open mind. The school leaves the discussion of the origins of life to

THE THEORY OF EVOLUTION IS A RESULT OF ERRONEOUS EXTRAPOLATION

> individual students and their families. As a standards-driven district, class instruction focuses upon preparing students to achieve proficiency on standards-based assessments.

For me, this was a fair compromise. Look, it didn't even require the students to study creationism. It only required the teachers to *mention* that there is an alternative hypothesis to evolution, which is creation. And that is a true statement, i.e., there is indeed an alternative hypothesis, so let us make this fact known to students!

The details of what happened next can be found in Wikipedia or Wikisource, so I will leave that to you; otherwise, this will get too long. I will just briefly summarize what happened next.

Basically, some of the pro-evolution board members resigned; the teachers did not want to read this statement; and the ACLU, on behalf of eleven parents, sued the board. So you got issues of free speech, separation of church and state, among many other things, coming into play. It was a long trial which took 40 days. There was even a book written about it. *40 Days and 40 Nights* by Matthew Chapman, a great-great-grandson of Charles Darwin.

Judge John E. Jones III issued the decision in the case and decided against the intelligent design proponents. His decision was 139 pages long, in which he wrote:

> The evidence at trial demonstrates that ID is nothing less than the progeny of creationism.
>
> The overwhelming evidence at trial established that ID is a religious view, a mere re-labeling of creationism, and not a scientific theory. (p. 31, 43)

So the judge decided in favor of evolution and against creation. Now, evolutionists might say "You see? It's been decided by a judge, so we are correct. Evolution is correct!"

Whoa! Wait! Stop!

Evolutionists are extrapolating again!

Just because the judge decided that Michael Behe was wrong doesn't mean Behe is wrong. I think Michael Behe is correct!

Are judges infallible? They are humans too!

How many innocent people have been convicted by judges? Will you let a judge decide matters of science, religion, and faith? Who are you to say that there is no God who created the universe and life itself, and you're saying that everything just happened by chance? Belief in evolution is a religion. It is not based on science. It is based on faith! There is no scientific basis for one species transforming to another, let alone life arising through random chances from non-living components. If we are teaching this religion in schools, why not teach all religions?

You might be thinking now, "Show me a proof there is God."

You can read ahead in later chapters where I discuss that in more detail. To briefly address that, I will say this: It is not a matter of opinion or even scientific evidence to decide whether or not God exists. Whether you like it or not, whether you believe or not, does not determine whether God exists or not. He exists. God created the universe and life itself. Creation is correct, and evolution is wrong.

What I'm simply doing is presenting an evidence to *help* you believe that there is God, and evolution is wrong. Not even science, mathematics, philosophy, your opinion, or anybody's opinion will be enough to decide once and for all if God exists or not. Even the most intelligent human being does not have enough brain power to use logic, reason, or even scientific method to answer and understand everything. Even mathematics fails in this regard. Imagine, mathematics failing in this regard? How much more of the other disciplines? Name it—astronomy, physics, chemistry, biology, law. This human "knowledge base" is not enough and will never be enough to match God's wisdom.

You can ask, "Why is God invisible?" "Why did God do this, do that?" "Why didn't God do it this way and that?" "Why can't God simply send an email message to everyone?" "Why is the world not perfect?" You can keep asking all the whys you can think of.

This is what the Bible says about that. Read Isaiah 55:8–9 (NIV), it says, "*For my thoughts are not your thoughts, neither are your*

THE THEORY OF EVOLUTION IS A RESULT OF ERRONEOUS EXTRAPOLATION

ways my ways, declares the Lord. As the heavens are higher than the earth, so are my ways higher than your ways and my thoughts than your thoughts."

If you understood the deep meaning of that, well and good. You are on the right path. If not, keep reading, perhaps there's hope.

You might ask, "So why did you write this book?" To help you see what most people don't see, even the most-learned man may not be able to understand this. Perhaps there is still hope for mankind to find his creator.

For now, I will defer this discussion to later chapters.

We need to look at mathematics, in particular, probability. You might ask, "I thought mathematics isn't enough?" Well, mathematics is the language I want to use to convey this message, and I'm the author.

CHAPTER 3

Mathematics, DNA, and Nanomachines

In 1990, I read a question on probability posted in "Ask Marilyn," a brain-teaser column in *Parade Magazine*, where readers would send in questions, and Marilyn would answer them. The column was by Marilyn Vos Savant. For background, Marilyn Vos Savant was listed in the *Guinness Book of World Records* as the person with the "World's Highest IQ."

The question one reader sent was, "Suppose you're on a game show, and you're given the choice of three doors. Behind one door is a car, behind the other doors are goats. You pick a door, say number 1. And the host, who knows what's behind the doors, opens another door, say number 3, which has a goat. He says to you, 'Do you want to pick door number 2?' Is it to your advantage to switch your choice of doors?"

Marilyn's response was, "Yes, you should switch. The first door has a 1/3 chance of winning, but the second door has 2/3 chance."

When I read this, I agreed with Marilyn's answer. I said to myself, "Of course, she's right. She's really smart."

Little did I know that that column, that short correct response from Marilyn would turn out to be one of the most heated statistical debates in recent history, even among "experts."

Marilyn received tens of thousands of letters in the ensuing days and weeks, many from noted mathematics professors, PhDs,

THE THEORY OF EVOLUTION IS A RESULT OF ERRONEOUS EXTRAPOLATION

and scholars telling her that she was wrong. Some are even MIT professors, renowned mathematicians and statisticians, and even MacArthur "Genius" Fellows.

She was even ridiculed in some of the responses. You can check her website and see for yourself: http://marilynvossavant.com/game-show-problem.

In the ensuing weeks, Marilyn kept saying that she's right. And she's right! But the letters kept coming in from mathematicians and PhDs saying she's wrong.

How can these bright people be stumped by a simple problem like this? They are supposed to be experts in this field.

Let's look at this problem deeper.

Say there are three letters—C (car), G (goat), and G (goat). If you pick a door (letter) by random guess, the probability that you'll pick C is 1/3, and the probability you'll pick a G is 2/3. That's how simple it is.

Now, let's say the host says, "I will open another door." Will that change your probability of picking C in your initial pick? No.

Now, the host is *about to* open another door. Will that change your probability of picking C? Again, no.

Now, the host opens one of the other two doors that you didn't pick and shows a goat. Has that changed your probability of picking C in your initial pick? Again, no.

It's because the host knows what are behind all the doors and can always pick another door with a goat.

In this case, the host has "intelligence" and actually, perhaps by "design," he wanted you to have a higher chance of winning the car by switching. So you should switch!

In your first pick, the prob (C) = 1/3. That's the probability of being right. So the probability of being wrong is 1−1/3 = 2/3.

Let's look at it this way. There are simply two "houses." House number 1 (containing the one door you picked) has 1/3 chance of containing the car. The other house, house number 2 (a house with the two doors you didn't pick) has 2/3 chance of containing the car.

House number 1 has one door, and house number 2 has two doors.

Now, one of house number 2's doors is opened, and it has a goat. Remember, house number 2 has 2/3 chance of containing the car, and house number 1 has 1/3 chance. Which *house* should you pick now? Of course, house number 2! And by doing so, your chance of picking the winning door is 2/3. By switching, you've increased your probability of success.

That's how simple it is. And yet intelligent people get stumped by this.

Let's put a little twist to this problem. Now there are 1,000,000 doors and still, only one car, but this car is divided into four parts: lights (L), internal engine (I), fuel tank (F), and everything else (E). And the game has now four stages. At each stage, your goal is to win a part by guessing where it is. You have to be correct in all stages in order to win the car!

I will call this new multistage game *I*ntegrated *N*ano *T*echnology based *A*dvanced *P*rotein *S*ynthesis or *INTAPS*.

Stage 1: The host tells you that only one of the doors has the part L (lights), the other 999,999 are all empty. You pick a door randomly. The host now opens 999,998 empty doors out of the 999,999 that you didn't pick. He asks, "Do you want to switch or stay with your pick?" Based on your strategy, you'll either win the part L and move on to the next stage or go home empty handed.

Stage 2: Same rules as stage 1 but you are trying to win part I (internal engine).

Stage 3. Same rules as stage 1 but now you're trying to win part F (fuel tank).

Stage 4. Same rules as stage 1 but now you're trying to win part E (everything else).

Now, you are given two strategies—E and C:
Strategy E: Do not switch at all stages and leave it to chance.
Strategy C: Switch at all stages.

THE THEORY OF EVOLUTION IS A RESULT OF ERRONEOUS EXTRAPOLATION

Let's now find out which strategy has the higher chance of winning the game.

Strategy E's probability of winning = 10^{-24}

= 1/1,000,000,000,000,000,000,000,000

This is because at every stage, you have 10^{-6} = 1/1,000,000 chance of picking the right door. Since there are four stages, you multiply this four times, and you get 10^{-24}.

Strategy C's probability of winning = $(999,999/1,000,000)^4$ = 99.99999999999999999 percent, in all practicality, you will win and assemble the parts for L-I-F-E.

The only way for this to work is if the host has the knowledge (intelligence) of which of the 1,000,000 doors contains the desired part at each stage. Also note that the host, by design, has put the right sequence of parts L, I, F, E at each stage so at the end, they will "self-assemble" and spell LIFE. Randomly guessing will not cut it. Did I mention self-assembly? We will get to that later.

By now, you might start to see the analogy of this game with the issue at hand.

Now, let's add another twist. The host now says, "I will *not* open new doors but from the one you pick, I can let you randomly 'mutate' at 1 percent rate. If you do mutate, I will let you choose a door from the other 999,999 unopened doors." Let's call this strategy A.

Strategy A: mutate at 1 percent then randomly choose

Do you think this will improve your chance of winning when compared to E? Yes, it will but not a lot.

Probability of winning using strategy A is calculated this way:

(Prob of guessing right in each stage) ^{4} = [Pr (of no mutation) * Pr (of guessing right given no mutation) + Pr (of mutation) * Pr (of guessing right given there is mutation)]^4

= $[(1-.01) * (1/10^6) + .01 * (1/999,999)^6]^4$

< 10^{-23}

Mutation helped but the chance is still very small. Actual observed rate of mutation at the DNA base-pair (bp) level is very small, in the magnitude $2 * 10^{-8}$ per base pair per replication event.

What I'm doing with strategy A is simulating the production of proteins. Think of the building blocks of the flagellum. Those are proteins, right? And the building blocks of proteins? Amino acids (AA). And how are amino acids formed? They are formed from the sequencing of the base pairs in the DNA. In reality, it is more complicated than this. As you can see, building proteins is not that easy if done randomly. And to add to the complication, the shape of the protein is very important for its function. It is like solving a three-dimensional jigsaw puzzle, say, a doughnut-shaped object from blocks of various shapes and they link via "magnets" (think of positive charge being attracted to negative charge). And this is happening at the molecular level!

One argument I've heard goes something like this: "No matter how complex it is, and no matter how small the chance that random mutations in the DNA would create the required parts of the flagellum in the right order, means nothing. The fact that it exists means it happened."

Whoa! Wait! Stop!

Whoever would think that way should get more understanding of mathematics. The fact that something exists does not mean it existed by chance! Even if you fully understand how it was built, it doesn't mean it was built by chance.

The easiest illustration I can think of is a standard example in basic statistics.

Let's say somebody shows you a coin. One side is head (H), the other side is tails (T). You play a game with that person. He tosses the coin. If it's H, you give him $1. If it's T, he gives you $1. You asked him if it's a fair coin. Here is how the conversation between you and him goes:

HE. What do you mean by fair?
YOU. That the outcome is purely random.
HE. I don't know the future outcomes, it must be random. To find out, let's do trial tosses. (*He tosses it*). Okay, it's H. (*He tosses it again*). Now, it's T. You see? It's 50/50, it's random. Ready?
YOU. Okay, let's play.

THE THEORY OF EVOLUTION IS A RESULT OF ERRONEOUS EXTRAPOLATION

You played the game 1,000 times and, in the end, you lost $900; and you won $100, with a net loss of $800.

Now, do you think the coin is fair?

If you say *no*, good you got it. If you say *yes,* please read the references in statistics at the end of this book.

If you say, "It doesn't matter, what happened already happened. The coin has two sides, so it is random."

Please ask somebody to explain it to you. If you don't have a friend who can explain, go to any college/university, look for their math/stat department and ask for help.

So you lost $800 to this guy. You asked him:

You. I couldn't believe I lost $800 to you. Are you sure it is a fair coin?
He. It's a fair coin. We tested it, right? I just got lucky.
You. Sure, sure. I should not have gambled. What's your name?
He. Just call me CD.

As you were walking away, CD approached another person. Let's say his name is JM.

Here is how the conversation goes between CD and JM:

CD. Hi, what's your name? Would you like to play a coin tossing game?
JM. I'm JM. Can I look at your coin?
CD hands over the coin to JM. JM pulls out a magnifying glass and inspects it.
JM. This looks like a regular penny. It even has something written on it, 'In God We Trust.' I want to take a closer look.
JM pulls out an electron microscope and investigates the inside of the coin.
JM. Wow, I see some nanomachines inside. Very tiny machines. The machines appear to be able to control the outcome of the toss. Wait, I see a super-advanced programming code controlling the machines. The machines are producing some tiny things, and they turn into different shapes. Wow, this is so cool. It's amaz-

ing. This can't happen by chance and whoever created it is not human.

C<small>D</small>. What? I didn't know that. I didn't know that coins have complex mechanisms inside.

J<small>M</small>. Well, now you know. Have a good day.

Think about the above illustration really hard. Now, picture this: CD = Charles Darwin. Coin = the cell.

If you need to see it in a way statisticians would say it, it goes like this. We have a null hypothesis Ho and an alternate hypothesis Ha.

Ho: The coin is fair.
Ha: The coin is not fair.

Define an outcome as having a H or T in a coin toss. By doing the experiment sufficiently enough number of times, we get statistically significant samples of the outcomes. If that coin is indeed fair, the results of the experiment should be such that the observed frequencies of H and T fall within a range of possibilities, with associated probabilities, as predicted by the distribution of outcomes coming from a process that follows true randomness. If the results deviate too much from that range, we should reject the null hypothesis. And we are sure of this decision at a high possibility of being correct because there are probabilities attached to the range. In this case, the results deviated too much from the range. So we reject the null hypothesis and conclude that the coin is not fair.

In the case of JM, he didn't even try the experiment. He first looked at the structure of the inside of the coin. By doing so, he was able to determine the right decision. In your case, you gambled and did the experiment. And you lost.

Let's now take a look inside the DNA. The DNA and RNA work together to make proteins in every cell. Let's list a few of its properties here.

THE THEORY OF EVOLUTION IS A RESULT OF ERRONEOUS EXTRAPOLATION

1. All life forms have DNA. An evolutionist might say, "You see, that's a proof of evolution. Every life form has DNA, so all life forms must have started from the same organism." Whoa! Wait! Stop! That's an erroneous extrapolation!
2. Every DNA is composed of base pairs.
    a) A base pair (BP) is one of these: A-T or G-C combination. A is Adenine, C is Cytosine, G is Guanine, and T is Thymine. These are chemicals that bond via hydrogen bonds. RNA has Uracil (U) too.
    b) In DNA, A always pairs with T and vice versa. G always pairs with C and vice versa. In RNA A pairs with U, G pairs with C.
        Think of the DNA as a set of rules and the RNA is an "interpreter."
    c) A codon is a sequence of 3 base pairs. All organisms use 3 base pairs in a codon. (An evolutionist might say that's a proof of evolution! Whoa! Wait! Stop! That's an extrapolation!)
    d) A codon is used to produce an amino acid (AA). There are 20 amino acids. Amino acids are strung together to form proteins.
    e) Proteins fold into 3-dimensional structures which determine their function.
        i. If there are < 50 AAs, it is called a peptide.
        ii. If there are 50–100 AAs, it is called a small protein.
        iii. An average protein has 300–400 AAs.
        iv. If it has > 500 AAs, it is called a large protein.
    f) A single AA error (mutation) can result in abnormal biological function.
        i. Example: Sickle cell anemia in humans, a crippling blood disorder is a result of a single cell mutation where valine (Val) is translated in place of a glutamine acid (Glu6).
3. To produce an amino acid, the DNA is read like a programming code where the RNA copies and translates the code. There are even *"stop"* sequences, just like in a pro-

gramming language. In programming languages, there are *if-then* statements. In DNA processing, there are also transcriptional activators and repressors that act like *if-then* statements. Amazing!
4. There are many resources you can find to get more details on this process. I have provided only a handful of properties here. Watch these two videos: https://youtu.be/00vBqYDBW5s and https://youtu.be/fpHaxzroYxg.
5. DNA replication is extremely accurate. The error is about 1 mistake in a billion nucleotides. It even has a "self-correcting" mechanism.

That's a high-level background. If you need more details, you can find many resources on the web and standard textbooks.

Now, let's take a closer look at the flagellum.

Recent research has revealed how complex this machine is. Look at the reference section for some links.

In the publication *The bacterial flagellar motor structure and function of a complex molecular machine* by Vonderviszt F. and Namba K., the authors said, "The bacterial flagellum is a biological macromolecular nanomachine for locomotion." They went on describing the complex molecular structure and assembly design of the various components of the flagellum. It's an outstanding research based on cutting-edge science.

In this article, the authors said:

> The well-established abilities of biological macromolecules to self-organize into complex three-dimensional architecture therefore give us a wonderful opportunity to use them in nano-technology applications such as design of self-assembling nanostructures for nano-devices and nano-systems.

THE THEORY OF EVOLUTION IS A RESULT OF ERRONEOUS EXTRAPOLATION

So researchers in nanotechnology are attempting to *reverse engineer* how these complex parts they see inside the cells are made, such as the flagellum.

Okay, so let's summarize in steps how proteins are made.

1. The DNA/RNA code acts like a programming language for creating amino acids.
2. Amino acids link to form protein peptides.
3. These proteins fold into different structures or shapes.
4. These shaped proteins are parts of a nanomachine with specific functions.

Proteins fold into three-dimensional structures which determine their function. This folding mechanism is at the atomic level based on the well-defined three-dimensional atomic arrangements of the atoms of the proteins.

Vonderviszt and Namba further said:

> The self-assembly process are all driven by precise recognition of template structures by assembling molecules, both of which have well-defined three-dimensional atomic arrangements for specific binding and interactions and yet extensive flexibility of these molecules including disordered conformation plays essential roles in formation and stabilization of complex structures that are otherwise impossible to form.

Note that last part: "otherwise impossible to form."

First, let's see how this can be reversed engineered.

1. Look at the protein structure of a part in a flagellum you are interested in.
2. Locate the precise atoms responsible for folding the protein at the exact location in three-dimension. This step is informational only at this point. The parts self-assemble!

3. Find those proteins and link them in the precise order and see if they really fold into the desired shape. To do this, get protein molecules from available sources (synthesize or get from organic specimens) and artificially link them together. If they fold into the correct shape, you're good. If not, try again.
4. Assume you are successful in step 3. Now, you know the right proteins. Get the amino acid sequence of these proteins. There are well-established protein sequencing techniques that you can use to do this, such as the Edman degradation process.
5. Find the code in the DNA that produced those amino acids. There are techniques for this (e.g. Sanger [dideoxy] method). More advanced methods exist.
6. Replicate that part of the DNA, isolate it, and test if they really do produce the amino acids you want in the right sequence. There are also techniques to do this.
7. See if the amino acids you created in step 6 are actually produced by your "copied DNA code."
8. Test if the amino acids link together into the desired protein peptide.
9. If so, test if the proteins you've synthesized fold into the desired protein structure.

If everything works, well and good. You can now synthesize the part. You can even start with step 4 if you can synthesize the amino acids then work back.

Have scientists actually reverse engineered this? Yes.

Watch the video here: https://youtu.be/uw0-MHI_248

So scientists have studied this tiny little super advanced nanoscale motor inside the bacteria. They can now explain how it works. Great! Good job!

Here's the problem. They extrapolate!

Just because they see it in an organism, they *assume* it "developed" through "evolution."

THE THEORY OF EVOLUTION IS A RESULT OF ERRONEOUS EXTRAPOLATION

Just because they can explain how it works even to its tiniest detail, they *assume* evolution is the cause of its existence.

Here is an example. An article appeared in *Advanced Science* (a Wiley-Blackwell publication), published online on June 25, 2015, titled, "A Delicate Nanoscale Motor Made by Nature—The Bacterial Flagellar Motor" by Ruidong Xue, Qi Ma, and Fan Bai.

Note what's in the title: "made by nature." This is misleading! It was not made by nature. It was made by God. This is what I said earlier. The word *nature* is part of the problem. People are thinking that just because they see it in animals, plants, on planet earth, inside a bacteria, they think it just happened by chance.

In it, the authors wrote in part 6 and I quote: "The bacterial flagellar motor is the pinnacle of *evolutionary* bionanotechnology: a self-assembling nanoscale electric rotary motor that performs at higher speed and with greater efficiency than any man-made machine." (Emphasis added)

Note the word in italics, "evolutionary." That's a huge extrapolation! They haven't proven that it was "evolution" that is responsible for its existence! And they're using it as if it were a fact.

Studying these nanomachines offer great benefits to the society. It advances our knowledge and perhaps this can be used for good purposes. But, *scientists should* stop *putting their bias into their research publications*. It misleads the public.

Another example is the title of this research paper *Evolution of higher torque in Campylobacter-type bacterial flagellar motors* by Bonnie Chaban, Izaak Coleman, and Morgan Beeby, published January 8, 2018, in Scientific Reports 8, article number: 97 (2018).

Note the word "evolution" in the title itself. It seems to imply that the "higher torque" evolved!

What they detected is variation in torque or force of the motor and correlated it with "motor phylogeny." If you'll look up "phylogeny" in *Merriam-Webster*, you will find this: Phy-log-e-ny (noun)—the *evolutionary* history of a kind of organism.

The dictionary is biased too! This is insane. The educational system, the dictionary, the scientists, they've been brainwashed!

They proposed a model for its "evolution." Basically, the appearance of the "rings" around the rod appears to be the "first" parts to appear in faster motors. I added to the quotes around "first" parts since they are implying that the parts "evolved" one by one.

Now, let's go back to step 2 above. The folding mechanism is very complex. It is dependent on the atomic structure. And when they're all completed, they get assembled into a flagellum. When scientists observe this, they call it "self-assembly."

Once the correct protein molecules are in place, they fold as if they know how to fold by themselves and snap into place (self-assembly). Amazing!

How does this really work? Imagine the three-dimensional protein prior to folding into its final shape.

The atoms around the contours attract other atoms nearby. If proteins don't fold the right way, they become inactive. They do not function as intended and sometimes even become toxic. Many allergies are caused by bad folding because the immune system is not able to produce anti-bodies for certain protein shapes (think of virus).

There are mathematical models that have been developed for the prediction of the length of the linear chain in primary and secondary structural development. But predicting the final shape is even more difficult. Imagine a long twine of string of proteins. There are so many configurations that you can shape it into. Without an algorithm or just randomly trying to come up with a shape and simply relying on the random distributions of the atoms, most likely you will get a ball-shaped protein all the time.

But the nanoparts of the flagellum *self-assemble*!

Let's take a closer look at this "self-assembly" property.

In the flagella assembly, you will find two rings, L ring and P ring, around the base of the rod which holds the rest of the tail, like bushings in a nut. You can do an image search for the fine details of it on the web and you'll see.

It is like a bushing or washer. And these rings are made up of different proteins, even different from the rod.

The probability that these three parts line up like this randomly is very small, practically zero. Just imagine putting two bushings and

THE THEORY OF EVOLUTION IS A RESULT OF ERRONEOUS EXTRAPOLATION

a screw the size of the hole of the bushings so they'll fit snuggly if you guide the screw into the bushings. Now disassemble them. Put the three parts in a metal tank and shake the tank as many times as you can. Do you think the screw will ever line up with the two bushings' holes and assemble themselves? Highly improbable.

This means the final shape of the proteins and their destination have already been "pre-determined" by the time the amino acids were being made. In other words, the whole sequence of producing the amino acid, the proteins, the folding of the proteins, and finally snapping into place (self-assembly) has been predetermined from the time the amino acids were created. Studies have shown that enzymes and other factors play a role too. But let's factor that out for now.

In order to get the right sequence of amino acids, you have to get the codons right, meaning the right sequence of base pairs.

In the next few sections, what I will do is demonstrate that the flagellum or any complex molecular nanomachine is irreducibly complex. Wow! You might ask, "Hasn't a judge decided that already?"

Uh, didn't I tell you already in chapter 2 that it is not your opinion or anybody's opinion, not even science, physics, or mathematics, or even law, will decide whether God exists or not? God exists.

What I'm doing is helping people *see* and open their eyes to the truth and stop being blind.

If I can convince you through mathematics that the flagellum did not evolve, and instead it is irreducibly complex, the alternative hypothesis is it was created. There is a creator.

Calculating probabilities when large numbers are involved tends to be difficult. So I used the concept of *upper bound*. An upper bound is a number that will not be exceeded by the "unknown" number you are trying to estimate. So for example, you are asked, in a deck of 52 cards, if I draw one card, what is the probability of drawing the ace of spades? If you answered, "I think it is not more than 1/50," you are correct. If you said, "not more than 1/40," that's correct too. It may be conservative or not conservative estimate depending on who the consumer of that estimate is. So among different estimates, there is the best estimate, a conservative estimate and so on. In this case, we are just looking for an estimate of the upper bound. The actual answer could be much less.

Now, let's look again at irreducibly complex. It says if one part is removed, the whole machine does not function anymore. All the parts have to be there, i.e., all or nothing.

Remember, the proteins in the nanomachines? They *self-assemble*. And I mentioned earlier, the AA's sequence has "predetermined" this self-assembly feature. It's like buying a jigsaw puzzle and the factory magnetized the edges. So when you opened the package, they self-assemble. Therefore, the "factory" already determined the solution, and they labeled the package *"No Assembly Required."*

By irreducibly complex machine, I will use the Jenga game as an analogy. Jenga, created by Lesli Scott, is a game of skill and strategy. Basically, there is a tower of 54 wooden blocks, 18 levels high with 3 blocks on each level, piled high in an orderly way with each level alternating in wood orientation direction. Each block is about 1/2" × 1" × 3". The blocks have small random variation and this creates some uncertainty in the game.

Two players play this by removing a block, any block, taking turns. If a player removes or touches a block and the tower crashes down, he loses. There is that one "final piece" that will make the tower collapse.

Now, evolutionists are claiming that the flagellum is not irreducibly complex. Let's play Jenga.

Research shows than even primitive bacteria had complex flagella already. If evolution were true, one of their ancestors must have none. Now, if evolutionists would say even the very first bacteria had flagella already, then we're done! The bacteria was created that way! It did not evolve.

For clarity of argument, assuming evolution were true, let's find that very first "hypothetical" bacteria with no tail. Let's add the parts for its flagellum via random mutation.

Let's calculate an *upper bound* of the probability of producing a new part. Now, we are to randomly produce it via mutation, right? Remember, our bacteria has "no tail"—no part at all.

If there is zero mutation, the bacteria will not have any chance of growing a tail. We need mutation for evolution to have any chance.

THE THEORY OF EVOLUTION IS A RESULT OF ERRONEOUS EXTRAPOLATION

First let's do a *bottom-up approach*: DNA to parts.

Assume we need a small protein for our first part, say 50 amino acids. Each AA is produced by a codon. Each codon has 3 base pairs. So we need 150 base pairs to produce 50 AAs.

The bacteria has 3,000,000 base pairs.

3,000,000/150 = 20,000 partitions (non-overlapping regions or sections). This is the minimum number of partitions of length 150.

Randomly guessing which 150-bp sequence to change = 1/20,000. Now, here's a complication. To get the right protein then make it fold to the right shape is a very complex process because you have to get the right AAs then link them in a way so that the 3-D structure is such that the atoms line up and then they self-assemble. Plus, it could be any group (any number of elements) out of the 20,000 that needs to change (mutate) to get the right protein and shape. There are $2^{20,000}$ possibilities just for the possible groups. For a reference number, $2^{1,000} = 10^{301}$. What more if we include the arrangements within those groups? So guessing which of these is the right one to mutate is almost an impossibility. We haven't even factored in the folding. If you're a nanotech engineer, it is like figuring out how to "program" the DNA itself! This approach is not looking good for evolution.

Top-down approach. We will look for a denominator to calculate our probability.

We're only looking for an upper bound. Let's look at it this way. There are only 20 AAs to choose from. Let's assume no repetitions for now. Including repetitions will increase the denominator. The number of possible groups from 20 is $2^{20} > 10^6$. Each group can have 1, 2, 3 . . . 20 elements. We can possibly weed out a few possibilities, like single AAs, but it will not reduce the possible number significantly. Once the group is formed, the order is important. So we have to permute each group in the 10^6 count! This means multiplying by n! each group of n. That's a really big number because $20! = 2.4*10^{18}$ already. This alone shows you that even if there is mutation, it is difficult to get the right protein. There may be no protein at all! Now, let's look at the middle number, 10. $10! = 3.6*10^6$. So, we know that the denominator we're looking for is *at least* this: the number of groups (10^6) * this number ($3.6*10^6$) = $3.6 * 10^{12}$. And note, I haven't considered the repetitions of AAs at this point.

Let's go through the calculations.

Expected number of mutations per bacterial replication = $2 * 10^{-8}$ (observed mutation rate per base pair) * $3*10^6$ (no. of base pairs) = $6* 10^{-2}$

Pr (mutate then guess which group of the 20 AAs to choose from then arrange) = $6 * 10^{-2} / 3.6 * 10^{12} = 1.6*10^{-14}$. So this number is an upper bound. Now, $1.6*10^{-14} < 10^{-13}$.

What's the significance of this number 10^{-13}? This is an *upper bound* for the probability of producing a new protein part through mutation. This is a *generous upper bound* because I didn't even consider the repetitions in the AAs and the folding process (the enzymes, temperature, acidity that I've factored out).

Like in Jenga, there is that "first" block which when you remove, the tower collapses. There is that minimal set of parts to even call it a machine. It can't be zero though.

Let N = minimum required parts to have a nanomachine. N > 0.

For now, assume N = 4 parts. Each part is one protein: rod, ring, motor, and power supply. Each is made of a unique protein. This is as simple as you can get.

Now, the flagellum needs a minimum of four simple parts or four unique proteins to even have a nanomachine. If these parts are already present in a bacteria, then they would self-assemble. And we have a flagellum.

Let's start with a bacteria with no flagellum but doesn't die and lived billions of years ago. Assume it mutates and produced the first part, a ring. The probability of this occurring is 10^{-13} (upper bound) using very generous assumptions. Assume the ring just floats inside the bacteria.

Assume it mutates again after one generation to produce the next part, say, the motor. The probability is 10^{-13} again.

Assume it mutates again to produce the next part, say, the rod. The probability is 10^{-13} again.

Then finally, it mutates to produce the power Supply. The probability upper bound is 10^{-13} again.

THE THEORY OF EVOLUTION IS A RESULT OF ERRONEOUS EXTRAPOLATION

These parts self-assemble to form the very first flagellum. The probability of this sequence of events happening is 10^{-52} * no. of permutations of 4 objects (since they can be produced in any order, i.e., rod could come before motor) = $4! * (10^{-13})^4 = 2.4 * 10^{-51}$ (i.e., $N!*p^N$, where p = pr of success)

Here is the key. If evolution were true (null hypothesis), this four-part flagellum would "evolve" into the present-day flagellum because we observe and know that modern-day flagella have more parts.

The present-day flagellum has about 25–40 different unique parts. Some have even more, and some parts are repeated. So we're not even counting here the number of repetitions of parts! Let's use the number 25 for the number of parts for demonstration. The probability that the next 21 will be produced is $21! * (10^{-13})^{21} = 5*10^{19} * 10^{-273} = 5*10^{-254}$.

This is the probability for an ancient 4-part flagellum to "evolve" into a 25-part flagellum.

Now, the probability of the whole process (from no tail) = prob(first 4) * prob (next 21) = $5 * 10^{-300}$

This is an upper bound.

If one argues like, maybe it's not 4, it's 10, then the upper bound is still in the order of 10^{-300}. This is because if we move N up or down, we get N initial parts but 25-N remaining parts. For N = 0, the term 10^{-325}, where 325 = 25 * 13, will be there and the factorial term, $25! = 1.55 * 10^{25}$.

If N= 10, this probability is:

$10! * (10^{-13})^{10} * 15! * (10^{-13})^{15} = 4.75 * 10^{-307}$

If N= 0, the probability is $25!*10^{-325} = 1.55 * 10^{-300}$

Evolutionists will now say, there's plenty of time and plenty of bacteria.

Let's check that out.

Bacteria multiply every half hour. So in a day, there are 48 generations. In a year, there will be 17,520 generations.

In 4 billion years (assumed age of earth, upper bound), there would be $7 * 10^{13}$ generations. So say if 10^{30} bacteria were in the initial population since earth began (an upper bound) and this is "stable" (every generation has this number), the expected number of bacteria with 25-part flagella today is: $1.55*10^{-300} * 7*10^{13} * 10^{30} = 10^{-257}$. Basically zero. This means if we were to rely only on random processes to produce the flagellum, we will get nothing.

What this calculation shows is if a 25-part flagella were to be produced through random mutations with 10^{30} initial bacterial population over 4 billion years, no flagella would exist today. *But we see 25-part flagella today!* They could not possibly have been manufactured through random processing, this means evolution is wrong.

Note, the 10^{-300} is an upper bound.

The 4 billion years is an upper bound.

Even if mutation rate is doubled, the probability increases by $2^{25} = 3*10^7$, still not enough.

If quadrupled, it increases by $9*10^{14}$, still not enough.

Just keep adding 7 to the power of 10 and so on.

Now, you might be thinking, "I can plug any big number here to prove evolution." It is not going to work like that. A random process follows some basic mathematical and statistical laws. Basically, if the rate of mutation deviates too far consistently from the known mean, it means it is not caused by random reasons. There would be outliers, yes, but the "error" from the expected mean and variances of a random process follow some distributions too.

How about survival rates, you might ask. The mutation rate is more important here. You can simulate and put in "assumptions" in the correlation or future survival rates but that will only put bias in your simulation. Here, I assumed a random, independent repeated trial experiment. To see this, say we have only one bacterium with no flagellum and overnight (by the act of God), all its DNA mutated so that next day the bacteria has a flagellum! That's not impossible (with God, nothing is impossible. Read Matthew 19:26). Yes, a bacteria with no tail can have a tail in a split second. But if that happens, it is not by random processes, it is an act of God.

Note, not all mutations are beneficial. Mutations can also kill the bacteria. So this rate has upper bound too! And if mutation rate changed enough to make this even possible, one would ask if that change in mutation rate is even random. Now, scientists have

THE THEORY OF EVOLUTION IS A RESULT OF ERRONEOUS EXTRAPOLATION

detected that the mutation rate in mtDNA (mitochondrial DNA) is higher than what's detected in the protein-producing process. By the way, bacteria don't have mitochondria.

And speaking of mtDNA, it actually acts like a *molecular clock*. It is because it is passed down only through your mother. Mutations passed down to male offspring are lost. However, not all mutations are passed down to the next generations. So scientists have studied the rate of mtDNA mutation in humans! Yes, humans. They can trace back in time man's roots. So if there really is an "Eve," the common ancestral mother of the human race, when did Eve walk on this earth? Some scientists estimated it at 6,500 years! Consistent with the Bible and not with evolution!

Now, back to irreducibly complex.

The 10^{30} is an estimate of the bacterial count in a "stable" population. This is based on estimates of present-day bacterial population on the whole planet! An *exaggerated* upper bound would be the number of *atoms* on earth. This is 10^{50}. Even if we plug in this number, it still wouldn't be enough.

(And if you want to include aliens, the number of atoms in the universe is 10^{80}. Still not enough.)

The theoretical maximum possible number of bacteria is 10^{80} (including aliens). With 4 billion years ($7*10^{13}$ generations) and 10^{-300} (upper bound) probability of producing a 25-part flagellum, we get no flagellum. This means, it is not possible to be produced via random processes. *The flagella could not have been produced via evolution!* Even using upper bounds of assumptions.

Let's summarize the arguments here via a Q&A between a creationist and an evolutionist:

Say the number of parts in a flagella that we observe today is 30 (nice round number). Here's the Q&A:

CREATIONIST. I claim this is irreducibly complex.
EVOLUTIONIST. No, it is not.
CREATIONIST. So where did the parts come from?
EVOLUTIONIST. They were just floating around.
CREATIONIST. That can't be because they self-assemble!

EVOLUTIONIST. Those floating around need an "extra part" to be produced by "evolution" in the future then when that extra part is produced, they will self-assemble.

CREATIONIST. How many were floating around before this final part is produced.

EVOLUTIONIST. Four (4).

CREATIONIST. Producing the first 4 has extremely low probability of occurrence, but it is still mathematically possible. However, we don't see any 4-part flagellum today. How can you explain the 30-parts we see now?

EVOLUTIONIST. They evolved gradually over time to 30.

CREATIONIST. The probability of going from 4 to 30 is so small. Then multiply that with the probability of having the 4 to begin with to having 30 makes it impossible even if we factor in elapsed time and initial population.

EVOLUTIONIST: Okay, so maybe it remained 4. Oops, that's not right because we see 30 now. Okay, maybe there were 29 floating around. Then the final 30th was produced then they all self-assembled.

CREATIONIST. If that's the case, it's irreducibly complex at 30. Because if you remove one part, it doesn't work. If it worked with 29 parts, then they self-assembled. The probability of producing those 29 randomly is very small, it is impossible given the time and population elements.

EVOLUTIONIST. No, no, no. It's not irreducibly complex at 29. Okay, so maybe it started with 28 parts floating around. Then the 29th part was produced via mutation. Then they still were floating around. Then the 30th part was produced, and they all self-assembled.

CREATIONIST. If that's the case, the probability that the first 28 parts were produced randomly is still very small. Now if you add the 29th and all 29 they self-assemble, so it means it's irreducibly complex at 29. If they don't self-assemble at 29, then it means its irreducibly complex at 30.

EVOLUTIONIST. What if I start with a different random set of 29?

THE THEORY OF EVOLUTION IS A RESULT OF ERRONEOUS EXTRAPOLATION

CREATIONIST. Then it's the same small probability. You can list all the possibilities and calculate backward. You'll see that no matter what random set of 29 you start with, you're back to the same problem.

EVOLUTIONIST. What if the 29 random parts self-assembled?

CREATIONIST. Then it's irreducibly complex at 29. Then you're back to the same problem again—calculating it forward or backward manner yields super small numbers for it to be feasible through random processes.

Note that the variables we used are mutation rate, number of parts, initial population, time, and an upper-bound calculation for the probability of randomly guessing the correct AAs (very generous), and the property of self-assembly. As can be seen, *it's the number of parts that's really important.*

We can now state the following:

Under the observed mutation rate of bacteria, any self-assembling nanomachine produced by the DNA inside a bacteria could not have been produced by random process such as evolution, given that the nanomachine has sufficiently large number of parts (N > 4).

If you'll go back to the calculations above, you will see that N = 5 is sufficiently large (given earth's age).

The key here is we are observing multi-part complex nanomachines with 30 plus parts!

They could not possibly have been produced through random process.

The evidence is what we observe here on earth! Given the analysis above, what it implies are these:

1. Earth is not old enough to produce these nanomachines by random processes.
2. There are not enough material on earth to make this even possible given the amount of material, the random processing (mutation rate and "evolution" if it were true), and time.

Assume, for example, we send humans (scientists) to another planet. Assume that these scientists have *no bias*—no preconceived belief, not brainwashed, no belief in how things are created, not even evolution, not even creation. Nothing. They're just intelligent, highly-trained humans.

After traveling a long distance, they reached a planet. In that planet, they found some machines (robots) made up of material found in that planet. So they investigated how these robots got produced in the first place.

They observed that there is a "process" that manufactures the robots' parts. Initially, they hypothesized that that process is a random process (null hypothesis). They said, "For these machines to be a product of random process, this planet should been very old, at least hundreds of billions of years old. And the amount of material needed for that random process to succeed is much bigger than the size of this planet."

Now, they calculated that the planet is only 4 billion years old. They concluded, "Some intelligent form created this."

The existence of these molecular nanomachines is not unique to bacteria. In fact, inside every living cell are many nanomachines performing specific functions. Take for instance the ribosome. The ribosome is a 50 plus-part nanomachine that produces the proteins. Following the probabilistic argument presented here, there is not enough time to produce the ribosome randomly. You might be asking 'Aren't there mathematical models that claim that there is enough time for evolution?". If you'll study those models carefully you'll see that these are incorrect models. They are circular claims, i.e., they assume that evolution is correct to prove evolution itself! Similar to phylogeny studies, they are all circular arguments. In science, logic and mathematics, you cannot assume that very thing you are trying to prove!

Now, evolutionists are claiming that all these complex features simply "appear" designed. This is a big blunder on their part. Mathematically, these complex nanomachines cannot be products of random processes. The *protein folding problem* is the question of how a protein's amino acid sequence determines its three-dimensional

THE THEORY OF EVOLUTION IS A RESULT OF ERRONEOUS EXTRAPOLATION

atomic structure. Scientists working in nanotechnology have made advances in solving the protein folding problem by reverse-engineering existing proteins and studying mathematical distributions of the observed conformations. But the question still remains – where did the DNA information code for the production of these proteins come from initially? Think about this: the DNA contains the code for protein production, the ribosome produces the proteins, the ribosome is made up of proteins, and the proteins are also producing the DNA! All these basic elements have to be present in the assumed "first" living cell! But the ribosome could not have been produced by random processes. It has too many parts!

This is what the Bible says about this:

> *"Hear this, you foolish and senseless people, who have eyes but do not see, who have ears but do not hear."* - Jeremiah 5:21

CHAPTER 4

God Exists

"In the beginning God created the heavens and the earth" (Gen. 1:1, NIV).

We can stop right here, there—first verse of the Bible. It explains everything. Very simple.

So how did the human race end up with so many theories, ideas, hypotheses about the origin of the universe, earth, and life itself?

You can research the history of all these theories—evolution, big bang, etc.—read about how scientists came up with those theories. But a question still lingers on why do humans look for alternative theories other than what is so simple and clear answer provided by the Bible?

I have a simple answer. Man does not want to believe in something that he cannot see, prove, or derive from scientific method.

Now, what does the Bible say about this? Again, I'll point you to Isaiah 55: 8–9 (NIV), which says, *"For my thoughts are not your thoughts, neither are your ways my ways, declares the Lord. As the heavens are higher than the earth, so are my ways higher than your ways and my thoughts than your thoughts."*

Science is no match to God's wisdom. Science can only help you "see" what God has created. Man can look inside the tiniest living cells or look deep into the outer vastness of the universe, and man will never ever be satisfied by what he can see—that there is a creator—because man refuses to believe.

THE THEORY OF EVOLUTION IS A RESULT OF ERRONEOUS EXTRAPOLATION

Evolutionists do not believe that the flagellum is irreducibly complex. It is irreducibly complex. I think the universe is irreducibly complex in a more general way. Scientists have observed that everything is so fine-tuned for life. Physicists have determined that there are universal constants governing our universe. But they are detecting that these constants are so perfectly calculated! If gravitational forces were just a tiny fraction stronger or weaker, the universe will collapse into itself or everything will disperse away. The same is true with other forces. You can say the same thing with our planet. Scientists are saying we are in a "goldilocks zone," where everything is just right. Our distance from the sun, the tilt of the earth, the magnetic field, the solar wind, even the speed of rotation, the size of the moon. These things did not happen by chance. God made all these things happen!

God is invisible. If you see a house, and you don't know who built it, you still believe that there was a builder, right?

> *"For every house is built by someone, but God is the builder of everything"* (Heb. 3:4, NIV).

So, why not apply the same reasoning when it comes to God? Why are you not using the same standard to deduce that there is God who created everything? Look at His creations, and you will know that He exists.

> *"The heavens declare the glory of God; the skies proclaim the work of His hands"* (Ps. 19:1, NIV).

Even mathematics is incomplete. Kurt Godel, a mathematician in the early part of 1900s, came up with two incompleteness theorems. Basically, these are theorems about the inherent limitations of every formal axiomatic system containing basic arithmetic. Arithmetic here includes the basic operations that we know but also the abstract forms of these. Godel's incompleteness theorems have two parts. The first theorem says that "no consistent system of axioms whose theorems can be listed by an effective procedure is capable of proving all the truths about the arithmetic of the natural numbers." The second incompleteness theorem states that "the system cannot demonstrate its own consistency" (Wikipedia). You can search for more details from other sources, but basically, Godel proved that there are things that we cannot prove using logic itself. There are many mathematical theorems related to this such as Tarski's theorem on undefinability of truth, and Turing's theorem that says there is no algorithm to solve the halting problem. Did I read that right? Mathematics says some truths cannot be defined let alone proven? Well, some say that this only applies to numbers, and they say that it is the reason why there are things in mathematics that cannot be proven, such a Goldbach's conjecture on prime numbers or even Riemann hypothesis on the roots of the zeta function.

I say mathematics, science, and logic are not enough to match God's wisdom.

And here we are, we have scientists, highly intelligent people, deciding that there is no God, as if it were the truth.

So how did this happen?

I had a conversation with an atheist once. It went like this (A = atheist, JM = me):

A. I don't believe in God.
Jm. Why?
A. The Bible is inconsistent.
Jm. Give me an example.
A. It says the earth is flat, but we know it is round.
Jm. Where in the Bible did you read that the earth is flat?
A. It talks about the "ends of the earth." Plural "ends." So earth has corners? No way that can be correct.
Jm. Let me help you. You cannot read the Bible like an ordinary book. It is hidden, it is not an open book. Only God's true messengers can teach from it. But let me tell you what I've been taught. The "ends" in "ends of the earth" signifies time, not direction, or corners. It is the *time* when the *end* of the world is near.

THE THEORY OF EVOLUTION IS A RESULT OF ERRONEOUS EXTRAPOLATION

A. That makes sense. You mean the Bible appears to have inconsistencies but in reality, they are not inconsistencies?

Jm. Yes, it is because of man's wrong interpretation of what's written. And the bad thing is, the Bible has been translated from many languages many times, and there are many incorrect translations.

A. Are there things in the Bible that are not hidden? Meaning we can read and easily understand?

Jm. Of course.

A. I'm a scientist. Are there scientific facts in the Bible?

Jm. Of course. But remember, the Bible is not a science book. It's man's guide to salvation. But let me give me you an example. If you'll read Job 26:7 (NIV), it says this: "He spreads out the north skies over empty space; he suspends the earth over nothing."

A. I thought the Bible was written thousands of years ago, how come it is talking about space?

Jm. Exactly!

A. Hey, tell me more about the "ends of the earth." You mean it is a time when we are close to the end of the world? When will it be?

Jm. Only God knows. Read Mark 13:32 (NIV), it says: "But about that day or hour no one knows, not even the angels in heaven, nor the Son, but only the Father."

A. Wow! I have a lot of questions. Can you help me answer them?

Jm. Come with me. You can ask any question you want.

A. Where?

Jm. To church.

A. I've been to a church before but the preacher was preaching his personal opinions and not what's in the Bible. I don't think he's the messenger you were talking about. He just wanted me to donate money to him.

Jm. Seek and you will find Him. Good luck, my friend.

"Above all, you must understand that in the last days scoffers will come, scoffing and following their own evil desires. They will say, "Where is this 'coming' he promised? Ever since our ancestors died, everything goes on as it has since the beginning of creation." But they deliberately forget that long ago by God's word the heavens came into being and the earth was formed out of water and by water. By these waters also the world of that time was deluged and destroyed. By the same word the present heavens and earth are reserved for fire, being kept for the day of judgment and destruction of the ungodly."

- II Peter 3:3-7 (NIV)

REFERENCES

1. Vonderviszt F., and K. Namba. "Structure, Function and Assembly of Flagellar Axial Proteins." https://www.ncbi.nlm.nih.gov/books/NBK6250/
2. Chaban, Bonnie, Izaak Coleman and Morgan Beeby. "Evolution of higher torque in *Campylobacter*-type bacterial flagellar motors." Published January 8, 2018 in Scientific Reports 8, Article number: 97 (2018). https://www.nature.com/articles/s41598-017-18115-1
3. Dubrovkin, A.M., R. Barille, E. Ortyl, and S. Zielinska. "Near-Field Optical control of Doughnut-Shaped Nanostructures."
4. Kojima, S., and Blair DF. "The bacterial flagellar motor: structure and function of a complex molecular machine." https://www.ncbi.nlm.nih.gov/pumed/15037363/
5. Berg, J., J. Tymoczko, and L. Stryer. 2002. *Biochemistry*. Fifth edition. New York: W.H. Freeman.
6. http://marilynvossavant.com/game-show-problem/
7. https://www.newgeology.us/presentation32.html
8. Osorio-Valeriano, Manuel, Javier de la Mora, Laura Camarena, and Georges Dreyfus. "Biochemical Characterization of the Flegellar Rod Components *Rhodobacter sphaeroides*: Properties and Interactions." *Journal of Bacteriology*. American Society for Microbiology.
9. http://blog.drwille.com/99-95-87-70-how-similar-is-the-human-genome-to-the-chimpanzee-genome/
10. Xue, Ruidong, Qi Ma, and Fan Bai. "A Delicate Nanoscale Motor Made by Nature—The Bacterial Flagellar Motor." *Advanced Science*. Wiley-Blackwell. Published online June 25, 2015.

11. Perdomo, Doranda, Melanie Bonhivers, and Derrick Robinson. "The Trypanosome Flagellar Pocket Collar and Its Ring Forming Protein tbBILB01." Published online on March 2, 2016.
12. https://www.ncbi.nlm.nih.gov/pmc/articles/PMC4810094/
13. https://evolutionnews.org/2011/03/michael_behe_hasnt_been_refute/
14. Brase and Brase. *Understandable Statistics*. Second edition.
15. Casella, George, and Roger Berger. *Statistical Inference*. Second edition.
16. https://nces.ed.gov/fastfacts, Organic Trade Association Surveys, US Dept. of Education.

ABOUT THE AUTHOR

Mr. Meimban is a quantitative analyst, mathematician and actuary with over thirty years of experience in derivative pricing, predictive analytics, risk management and software engineering. He plays golf, table tennis and practices jiujitsu. Mr. Meimban has a Master of Science in Mathematics degree from Florida State University. His current research initiative is in the modeling of rare events. His mission is to help bring people back to God.

CPSIA information can be obtained
at www.ICGtesting.com
Printed in the USA
BVHW071239280119
538842BV00004B/670/P